OUT OF PRINT
HELD AT
LAST COPY STORE

12/04/21

HAMPSHIRE COUNTY LIBRARY

WITHDRAWN.

CL.69(02/06.)

107

C013859584

Local Author

This book is due for return on or before the last date shown above: it may, subject to the book not being reserved by another reader, be renewed by personal application, post, or telephone, quoting this date and details of the book.

HAMPSHIRE COUNTY COUNCIL
County Library

100% recycled paper

BUTTERFLY LANDSCAPES

A CELEBRATION OF BRITISH BUTTERFLIES PAINTED IN NATURAL HABITAT

RICHARD TRATT

Langford Press

© Richard Tratt 2005

Langford Press
www.langford-press.co.uk

The rights of Richard Tratt to be identified as author and illustrator of this work
have been asserted in accordance with the Copyright, Designs and Patents act 1988

Printed and bound in Singapore under the supervision of
MRM Graphics Ltd. Winslow, Bucks. UK.

A CIP Record for this book is available from the British Library

ISBN: 1-904078-15-X

Dedicated to my loving wife, Hilary, who
has accompanied me on many butterfly adventures
and who has tirelessly given her help and support in
producing this book.

CONTENTS

	Introduction
1	Hillsides and Downland
17	Forest and Woodland Edges
35	Skippers
53	Swallowtail, Whites and Yellows
71	Hairstreaks
83	Coppers, Blues and Duke of Burgundy Fritillary
107	Emperors, White Admirals and Vanessids
123	Fritillaries
141	Browns
165	Small Butterflies in a large Landscape
185	Wild Meadows and Grassland
197	Garden Visitors
204	Rare Migrants and Extinct British Species
206	Species Index

Oil   36 × 40 in

BUTTERFLIES OF THE OPEN HEDGEROW

COLLECTION OF THE NEW FOREST MUSEUM, LYNDHURST

FOREWORD

The marvellous paintings in this book capture the reasons for my deep interest in butterflies from a young age. By good fortune, this has led to a lifetime career of butterfly study and conservation. These paintings echo the warmth and peace of warm summer days, and of course the dazzling beauty of the butterflies that bring different habitats to life. This is one of the main reasons why we should conserve these fantastic insects, so that future generations can enjoy that same experience - one that makes life worth living.

Richard Tratt has a unique gift in portraying butterflies and their habitats as a seamless entity, an integral part of the landscape. The book truly is a celebration of British Butterflies and I hope they bring you fresh inspiration and joy.

I am delighted that the publishers have generously offered to donate 5% ((£1.75) to Butterfly Conservation for every copy sold. Sadly our surveys have shown that 7 out of every 10 British species are declining and many are threatened. The money will therefore be used to support our crucial work to save species from extinction and bring about an environment rich in butterflies and moths.

Martin Warren

Dr Martin Warren
Chief Executive, Butterfly Conservation

If you would like to join butterfly Conservation or donate further towards our conservation work, or simply find out more about these fascinating insects, please visit our website:

www.butterfly-conservation.org

Or write to:
Butterfly Conservation, Manor Yard, East Lulworth, Dorset DT2 7PS

Saving butterflies, moths and their habitats

INTRODUCTION

As a young boy I developed an overwhelming passion for the natural World and a particular fascination for butterflies. Brilliant colours and intricate patterns were just a part of the attraction. I soon learned how each species has a distinctive flight and a very definite habitat preference. Watching butterflies has led me to wonderful wild places, warm woodland glades, flower rich downland, open moorland and sunny grassland slopes.

Compared to the burnt out look of the Mediterranean, Britain is a truly green country, everywhere lush with vegetation. After returning from a few trips abroad, I couldn't help but feel a new fondness for our damper and cooler climate. The landscape of the British Isles has a wonderful diversity of character, changing dramatically from county to county. In the East, from the white cliffs of Dover to the fens of East Anglia, and Northward to the Yorkshire Dales and the Scottish Highlands. In the West, from the heathlands of Dartmoor and the peat bogs of Ireland, to damp Snowdonia Valleys and Northward to the Lake District and Scotland's beautiful West Coast islands.

In this book, I've combined a lifetime career as a landscape painter, with a life long interest in butterflies. I've tried to capture the exciting and often fleeting moment of chancing upon each species. A new location, the mood of the day, and the warmth of sunshine are equal parts of the butterfly watching experience. Every year the situation is different, the numbers of each species varying considerably. As an example , the Holly Blue can sometimes be seen in every garden, yet the following year it becomes a rare sight.

I must admit that as a young boy I used to collect butterflies, yes the pin and the glass showcase! Back then nobody questioned it, my father encouraged me and I used to meet similar minded collectors out in the field. It stopped abruptly when I saw the error of my ways and persuaded my father to buy me a decent camera. This was the 1960's, the end of an age of plenty when small plots of unused and overgrown land were abundant and field edges were allowed to grow wild. Since then a great clean up has taken place, not only in all towns and villages but also in general agriculture .

Intensively farmed areas and chemical sprayed fields are no more than a desert for most butterflies. Land has become valuable and every house building opportunity taken. In residential areas, overgrown plots supporting Nettles, Thistles, Brambles and a mix of grasses and wildflowers, are seen by many as an eyesore, to be cleaned up by the owner or the council as soon as possible. People often ask, "where have all the butterflies gone?". The answer is very simple…The weeds have been removed! Yes, our native plants, known by tidy gardeners as weeds, really are important. Learn to love them! Otherwise the general decline in butterfly numbers will continue. Luckily a new move towards preservation has begun and a renewed interest and understanding of wild habitats

Butterfly Conservation was first formed in 1968 as "The British Butterfly Conservation Society". It's aim was to monitor and check the decline of so many British species over the previous fifty years. Since then, "Butterfly Conservation" has grown and flourished, with a positive and successful strategy for rebuilding stronger populations and reversing the species decline. A wide range of new and exciting knowledge has been gained about specific breeding requirements for rarer species. This has been used to great effect in re-establishing new colonies and rejuvenating existing ones.

Over the last hundred and fifty years, four of our resident species have become extinct. The Large Blue is a fifth casualty, but has been reintroduced, with great success, despite it's complex life cycle. (see page 102). Very often, an unmanaged habitat simply becomes "shaded out" by the rapid growth of scrub and tall bushes. When these are cleared, the wildflowers begin to recover, and within a few seasons, so do the butterflies. In many forest areas, edges of rides can be initially plentiful with species, but soon become unsuitable as the tree height increases. By widening these verges, the sunlight floods back into the wood and the habitat is restored.

Butterflies are the "canaries in the coalmine", their presence or absence providing a clear indicator for the natural health of our environment.

Butterflies are sun lovers, like myself, and many of my most treasured memories involve summer walks over butterfly territory. Chalk downland slopes are probably my favourite of all habitats. By midsummer, Marjoram and wild Thyme scent the air while scabious and knapweed cover the south facing slopes in a perfect blend of colour. Life doesn't get any better than this…..time to begin the next painting.

I'd like to express my gratitude to members of "Butterfly Conservation" and to all Nature Reserve wardens, who put their time and energy into maintaining these butterfly habitats, which I so love to paint.

A final word of thanks to my publisher, Ian Langford, who provided the opportunity and encouraged me to bring together a lifetime of paintings, to create "Butterfly Landscapes"

HILLSIDES AND DOWNLAND

HAMPSHIRE DOWNLAND

Oil   18 × 24 in

BUTTERFLIES OF THE SOUTHERN CHALK DOWNLANDS   (MARTIN DOWN)

Oil   20 × 40 in

COLLECTION OF THE NATURE IN ART MUSEUM, GLOUCESTER

The Southern half of Martin Down has been unploughed for centuries and is a haven for rare plants and butterflies. Higher up, along an ancient earthwork, Bockerley Dyke, managed grazing has kept the turf short to encourage demanding species, such as Adonis Blue and Silver spotted skipper. Elsewhere on this gently sloping reserve, are vast populations of Dark Green Fritillary, Chalkhill Blue and Marbled White. This painting depicts a hot afternoon in July, when the Down is at its flowering peak and the scent of midsummer fills the air.

BROWN ARGUS

ADONIS BLUE

CHALKHILL BLUE

COMMON BLUE

RED ADMIRALS AT HOMINGTON DOWN

Oil 20 x 43 in

Oil 16 × 12 in

DOWNLAND BUTTERFLIES

EVENING SHADOWS, MARTIN DOWN — Oil 12 × 10 in

BUTTERFLIES AT MAGDALEN HILL DOWN

Oil 22 × 36 in

In 1989 Butterfly Conservation began management of an overgrown hillside, Hampshire's Magdalen Hill Down. Scrub clearance, grazing and the introduction of wild flower seed from other chalk downland reserves, have restored this wonderful site, which now supports over 30 species of butterfly.

Butser Hill is a national nature reserve, the highest point of the South Downs and part of the Queen Elizabeth Country Park. In midsummer the downland slopes are a colourful patchwork of wild flowers. The chalk soil, along with regular grazing, encourages a wealth of low growing plants and fine grasses. Butterflies simply love purple flowers and here, Common and Chalkhill Blues feast upon Marjoram, while a Marbled White nectars nearby on Scabious.

BLUES AT BUTSER HILL — Oil 18 × 24 in

CHALK DOWNLAND, COOMBE BISSET

Oil 16 × 30 in

BEACON HILL NATURE RESERVE

Oil   16 × 24 in

Beacon Hill is a National Nature Reserve at the western end of the South Downs. It has a wonderful open outlook with panoramic views of the surrounding countryside. Beech woodland creates a wind break along the top of the warm furrowed slopes. The yellow flowers of Mignonette and Hawkweed jostle with the purples and blues of Marjoram, Knapweed and Scabious, deep rooted plants, enabling them to survive on the driest soils. This managed reserve is home to the rare Silver-spotted Skipper, shown here with Marbled White, Chalkhill Blue and Meadow Browns.

Acrylic   18 × 18 in

WOODLAND BUTTERFLIES

FOREST AND WOODLAND EDGES

THE BRAMBLE VISITOR

Oil   12 × 16 in

Brambles provide a very important nectar source for many species of butterfly, lining the edges of woods and hedgerows throughout the country. By June the buds open to reveal a mass of simple five petalled flowers, rising up above the sprawling plants. "Bramble Patch" was inspired by a visit to Bookham Common, Surrey, where the woodland glades provide an ideal haunt for the magnificent Silver-washed Fritillary, the focus of this painting. In July and August the blackberry bushes at the glade edge are crowded with butterflies. Set against the dark shadow of the wood, this painting emphasizes the sunlit foreground, also enjoyed by a Speckled Wood, Small Tortoiseshell and Small Whites.

BRAMBLE PATCH

COLLECTION OF THE NATURE IN ART MUSEUM, GLOUCESTER

Acrylic 18 × 24 in

Oil 15 × 16 in

WHITE ADMIRALS AT ASHURST WOOD

PURPLE EMPEROR Oil 16 × 12 in

PEARL-BORDERED FRITILLARIES IN WEST SCOTLAND

Oil 16 × 30 in

PREPARATORY SKETCHES FOR HAUNT OF THE RED ADMIRAL

HAUNT OF THE RED ADMIRAL   Oil   36 × 40 in

The painting, "Haunt of the White Admiral" takes me back to my first encounter with this striking butterfly. I was on a childhood holiday to the New Forest and caught sight of one at the edge of a woodland glade. It was a magnificent freshly emerged specimen with velvety black wings and contrasting white band. It seemed to jump out against the foliage. Since then I've been inspired to paint many White Admiral paintings. This is the largest so far, being 30 x 40 inches. A Small White and a Large Skipper share the same Thistle plant, while a Gatekeeper and Purple Hairstreak sit nearby.

In the painting "White Admirals at Ashurst Wood" (page 20) a pair can be seen nectaring on Bramble, at the edge of a woodland ride. Here, in the East of the New Forest, conditions have been recreated to encourage this butterfly to thrive once again, along with another spectacular woodland species, the Silver-washed Fritillary.

HAUNT OF THE WHITE ADMIRAL Acrylic 30 × 40 in

The butterfly season begins in early spring with the appearance of a few hibernating species on warm days. Seeing the Brimstone never fails to excite. The bright lemon yellow simply jumps out from the muted tones of late winter. This long lived butterfly hibernates in an outdoor situation, deep in a Holly bush or Ivy thicket. With wings closed it's unusual pointed shape acts as a camouflage amongst these evergreen leaves. After their long winter rest, the Brimstones have to gain strength, feeding up on early blossom. I enjoy seeing them nectaring on the striking flower clusters of Blackthorn, the inspiration for this painting. In search of a mate, a male and female circle around each other in courtship flight.

FIRST SIGN OF SPRING, THE BRIMSTONE — Oil 14 × 18 in

A QUIET CORNER

Oil   18 × 24 in

Sometimes while out walking, I'll pass a field corner along a woodland edge, where the wind suddenly drops to a murmur and the warmth of the sun's rays can truly be felt. In these secluded retreats, butterflies congregate, enjoying the shelter and the abundance of wild vegetation. Just such a place near my home inspired me to paint "A Quiet Corner", thick with Meadow Sweet and the scent of high summer. On a seed-head perch a Small Tortoiseshell basks in the sun, while a lazy Ringlet takes a break on the tip of a rusty dock stem. A Small White passes by, heading for a patch of flowering thistles.

The year 2004 was a exceptional year for the Silver-washed Fritillary. On an August visit to Bentley Wood, they could be seen nectaring upon every Thistle and Bramble patch. Hilary and I were able to watch their unique courtship flights and later, chanced upon a number of mating pairs along the edge of a shady track. In this painting I've tried to convey their subtle camouflage amongst the leafy surroundings. Vulnerable as a joined pair, they find a quiet hide-away, their silvery bands and delicate green colouration breaking up their outline.

MATING PAIR, SILVER-WASHED FRITILLARIES   Oil   12 x 16 in

SKIPPERS
HESPERIIDAE

This is a family of small moth-like butterflies, named after their skipping and dancing flight. When basking in the sunshine they have a characteristic way of holding their forewings at a 45 degree angle, while the hind wings remain flat. We have eight species in Britain. Six are golden brown and rely on grassland habitat for their life cycle. The remaining two, the Grizzled and Dingy Skippers, are a sub family, their larvae feeding on specific low growing plants rather than grasses.

CHEQUERED SKIPPER	*CARTEROCEPHALUS PALAEMON*
SMALL SKIPPER	*THYMELICUS SYLVESTRI*
ESSEX SKIPPER	*THYMELICUS LINEOLA*
LULWORTH SKIPPER	*THYMELICUS ACTEON*
LARGE SKIPPER	*OCHLODES VENATA*
SILVER-SPOTTED SKIPPER	*HESPERIA COMMA*
DINGY SKIPPER	*ERYNNIS TAGES*
GRIZZLED SKIPPER	*PYRGUS MALVAE*

CHEQUERED SKIPPER

RESTING POSITIONS - CHEQUERED SKIPPER

Always a rare butterfly, the Chequered Skipper was previously found in England, but unfortunately suffered a rapid decline from its woodland localities in the East Midlands. By the mid 1970's it was presumed to be extinct in this region. Luckily, a remaining colony was known to exist near Fort William and recently, other overlooked colonies have been discovered in this small area of West Scotland. Here, I've found it in woodland clearings and heathy sheltered valleys, but always in the vicinity of damp ground. This species looks drab when in flight, but is a distinctive bright little butterfly when approached closely. The male and female have very similar markings.

Flight Season :   mid May - June

Caterpillar's Foodplant :   Purple Moor Grass

CHEQUERED SKIPPER

WEST SCOTLAND HABITAT - THE CHEQUERED SKIPPER

SMALL SKIPPER

GRASSLAND HAVEN - SMALL SKIPPERS

Absent from Ireland, Scotland and Northern England, but otherwise common in most places where tall grasses are abundant. Habitats vary from woodland rides to roadside verges and rough open fields. The females are plain golden brown and the male is distinguished by a black scent line on the upper forewing. Sometimes found in vast colonies on waste ground or extensive rough grassland.

Flight Season : July and August

Caterpillar's Foodplant : Tall grasses with a preference for Yorkshire Fog

SMALL SKIPPER

SMALL SKIPPERS IN GRASSLAND HABITAT

ESSEX SKIPPER

SKIPPERS ON THE ESSEX MARSHES

Originally discovered in Essex this butterfly is now known to occur in most counties of the South East and the East Midlands. Scattered colonies occur as far west as Devon and Cornwall, mainly on a few coastal sites. There are two key differences between this and the Small Skipper, but both are very slight. Firstly, the black scent mark of the male is reduced to a thin black line running parallel to the upper edge of the wing. Secondly the tips of the antennae are jet black. Salt marshes of the east coast are a stronghold for this species, but further inland it enjoys the same locations as the Small Skipper.

Flight Season : July - August

Caterpillar's Foodplant : Various grasses -Cock's Foot, Couch grass, timothy grass

ESSEX SKIPPER

ESSEX SKIPPERS ALONG THE WOODLAND EDGE

LULWORTH SKIPPER

ALONG THE DORSET COAST PATH

The only resident butterfly named after a specific town, this skipper does indeed live in the Dorset Coast region either side of Lulworth Cove. It's range extends from Swanage to Weymouth. Here and at a couple of coastal sites in Devon and Cornwall, it remains fairly abundant and can be found sitting amongst tall grasses in sheltered areas of coastal cliffs and landslips. It has a fast flight and a darker colour than the common Small Skipper, so it is easily overlooked.

Flight Season : July and August

Caterpillar's Foodplant : Tor Grass

LULWORTH SKIPPER

LULWORTH SKIPPERS ABOVE THE COVE

LARGE SKIPPER

THISTLE FLOWERS AND SKIPPERS

The largest skipper, this is a common butterfly throughout the south, the midlands and Wales. More local in Northern England and southwest Scotland; absent from Ireland. It's preferred haunts are woodland edges and sheltered, sunny areas where tall grasses grow thickly. Often seen basking with open wings, or nectaring from favourite sources, thistle and bramble. It frequently chooses a leaf or a grass blade from which it darts off with a fast whirring flight, only to return to the same point again and again.. The upperside of the Large Skipper is very similar to the Silver-spotted Skipper but this species has a fairly plain underside and a different habitat.

Flight Season :   June - August

Caterpillar's Foodplant :   Cocks foot grass and other tall grasses.

LARGE SKIPPER

LARGE SKIPPERS IN A WOODLAND RIDE

SILVER-SPOTTED SKIPPER

SKIPPERS ON THE SOUTH DOWNS

A rarity, restricted to a few chalkhill localities in the South and south east. Found only on warm south facing slopes with a sparse covering of short turf. The sites I've visited have been grazed short by rabbits, to a point where patches of bare ground are exposed. Although very local, colonies may contain several hundred individuals. Similar in size and colour to the Large Skipper, but easily identified by its conspicuous silver marks on the underside.

Flight Season : August - mid September

Caterpillar's Foodplant : Fine Sheep's Fescue Grass

SILVER-SPOTTED SKIPPER

ON A CHALK HILLSIDE

DINGY SKIPPER

RESTING POSITIONS, THE DINGY SKIPPER

  I've found this moth like butterfly in many different habitats, the preference being for low vegetation and patches of bare ground, where it can spread it's wings and bask in warm sunshine. Like other skippers it has a rapid flight and is very difficult to follow. The wings soon become worn and develop a paler shade of brown. When resting, it often holds it's wings like a moth, as shown above. Commonest in the South and around the coastline of Wales and Northern England. A few colonies exist in North East Scotland and this is the only skipper to be found in Ireland.

  Flight Season :   May - June

  Caterpillar's Foodplant : Bird's Foot Trefoil

DINGY SKIPPER

BASKING ON THE DUNES

GRIZZLED SKIPPER

SUNNING POSITION - THE HILLSIDE TRACK

A tiny butterfly, often overlooked due to it's fast flight and size. Very attractive, especially on freshly emerged specimens, when the very dark wings contrast sharply with the white chequer pattern. The markings extend into the white fringes to give a distinctive edge to the wings. I usually see this skipper on downland, sunning itself on bare ground and less frequently on farmland and in woodland clearings. Often seen flying alongside the Dingy Skipper, The Grizzled Skipper's stronghold lies in the mid South and South Eastern counties, but it is becoming increasingly local. Elsewhere, colonies are confined to scattered coastal locations and a few larger woods in the Midland and Eastern counties.

Flight Season :   May - June

Caterpillar's Foodplant :   Wild Strawberry, Cinquefoil

GRIZZLED SKIPPER

THE GRIZZLED SKIPPER ON WILD STRAWBERRY

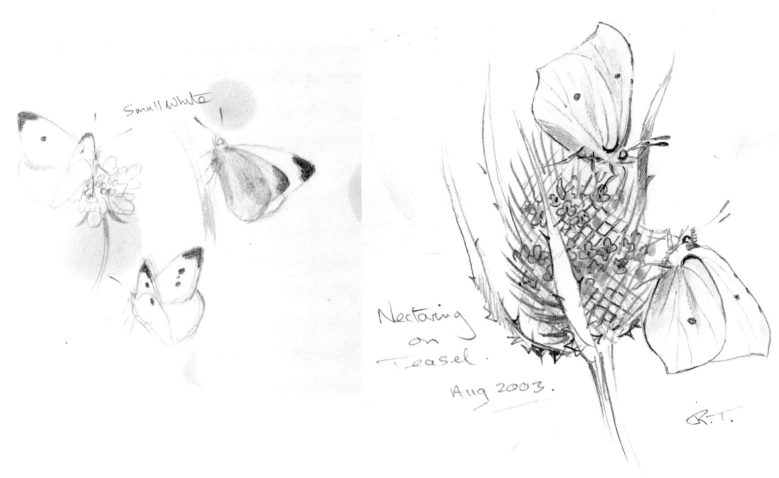

Small White

Nectaring on Teasel.
Aug 2003.
R.T.

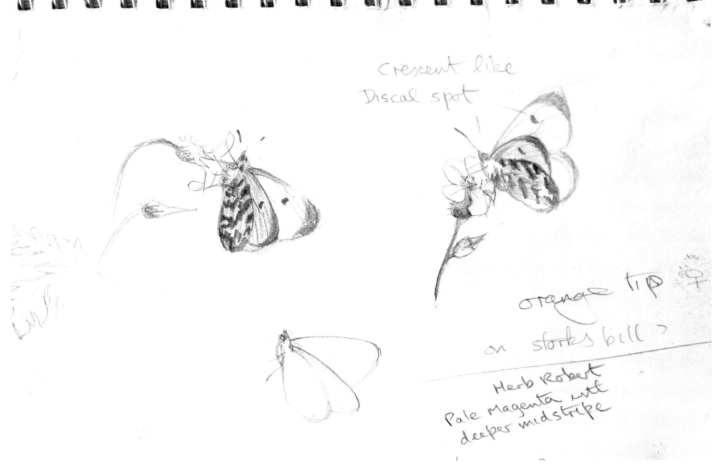

Crescent like Discal spot

Orange tip on storks bill?

Herb Robert
Pale magenta with deeper midstripe

SWALLOWTAIL

PAPILIONIDAE

This family of majestic butterflies includes the Apollos and the Birdwings. We have only one species, the Swallowtail, Britain's largest butterfly

SWALLOWTAIL *PAPILIO MACHAON*

WHITES AND YELLOWS

PIERIDAE

These butterflies are immediately recognised, even in flight for having a predominantly white or yellow wing colour. Apart from the Wood White, all species are wanderers, not living in colonies, but roaming freely for large distances. This habit takes them into gardens where they will nectar on a wide variety of flowers. Most can be encouraged to breed in gardens by introducing their correct food plant. The large and small whites don't need any encouragement!

SMALL WHITE *PIERIS RAPAE*
GREEN-VEINED WHITE *PIERIS NAPI*
LARGE WHITE *PIERIS BRASSICAE*
WOOD WHITE *LEPTIDEA SINAPIS*
ORANGE TIP *ANTHOCHARIS CARDAMIES*
CLOUDED YELLOW *COLIAS CROCEUS*
BRIMSTONE *GONEPTERYX RHAMNI*

SWALLOWTAIL

FLIGHT OVER THE REEDBEDS.

The Fens of East Anglia are Britain's only locality for this magnificent butterfly. It's range is limited further by the rarity of the caterpillar's foodplant, Milk Parsley. The Swallowtail is our largest resident species and is unmistakeable, with it's strong purposeful flight over the reed beds. I have seen this wonderful butterfly in quite large numbers at managed Nature Reserves in East Norfolk, it's main stronghold.

Flight Season :   Late May - mid July

Caterpillar's foodplant :   Milk-Parsley

SWALLOWTAIL

SWALLOWTAILS, NECTARING ON RAGGED ROBIN

SMALL WHITE

FEASTING ON EUPATORIUM

From early in the year to the first frosts, this butterfly is a familiar sight everywhere. The Small White is easily recognised by it's size and plain yellow underside. It is probably the most common garden visitor, nectaring on a wide variety of flowers. It frequents all types of habitat and existing populations are increased by incoming migrations from abroad. In gardens the small blue green velvet caterpillar is often found on cabbages or nasturtiums.

Flight Season : First brood - April - June Second brood - July - October

Caterpillar's foodplant : Brassicas, nasturtiums, wild crucifers

SMALL WHITE

SIMPLICITY, THE MALE SMALL WHITE

GREEN-VEINED WHITE

IN DAPPLED SHADE, GREEN-VEINED WHITES

Similar to the small white and easily mistaken in flight. However, when settled, the Green Veined White is seen to have a distinctive underside - as it's name suggests. This butterfly has a preference for damp ground and lush vegetation, but is also seen in gardens, lanes and woodland edges. Common and well distributed throughout the British Isles.

Flight Season : Mid April - June Second Brood - July - September

Caterpillar's Foodplant : Charlock, Hedge Mustard, Lady's Smock, Crucifers

GREEN-VEINED WHITE

THE WILD ROADSIDE WITH GREEN-VEINED WHITE

LARGE WHITE

SCABIOUS VISITOR

    This big distinctive butterfly is none other than the infamous 'Cabbage White'. The sleek and distinctive markings of the male are in contrast to the spotted wings of the larger heavier female. It has a strong fluttering flight and is a very common sight in gardens and allotments. The Large White is found throughout Britain and in some years, the population is boosted by vast migrations from the Continent.

Flight Season :   April - June   second brood - July - October

Caterpillar's Foodplant :   Brassicas, Crucifers and nasturtiums

LARGE WHITE

LAVENDER BORDER WITH LARGE WHITES

WOOD WHITE

WOOD WHITES AND FOODPLANT, MEADOW VETCHLING.

A very local species with a distinctive oval wing shape. Restricted to the Southern half of England and Wales, but is fairly common throughout Ireland. This is a delicate butterfly and I've always seen it in sunny open areas where trees and bushes provide protection from the wind. It has a weak slow flight and rests with wings closed. Recently, the Irish race, which prefers a more open habitat, has been identified as a separate species; the Real's Wood White, though differences in appearance are very slight.

Flight Season : May - June,

Caterpillar foodplants : Vetches, including Bird's-foot Trefoil, Meadow Vetchling and Tufted Vetch.

WOOD WHITE

WOOD WHITE AND LARGE SKIPPERS

ORANGE TIP

THE MALE AND FEMALE ORANGE TIP

Always a welcome sight after the long winter months, the Orange Tip is one of the earliest butterflies to emerge each spring. It can be seen along lanes, hedgerows and in gardens. However it's favourite habitat is in damp flower rich meadows, where Lady's Smock is abundant. The female lacks the orange tips and is most easily identified when settled. Common throughout Britain except for large areas of Central and West Scotland.

Flight Season : mid April - mid July

Caterpillar's Foodplant : Lady's Smock and Garlic Mustard.

ORANGE TIP

ORANGE TIPS ON LADY'S SMOCK

CLOUDED YELLOW

CLOUDED YELLOWS IN OCTOBER, BALLARD DOWN

Very distinctive, being Britain's only deep yellow butterfly. Can be quite rare in some years, with very few recorded, though occasionally, this well known migrant arrives on our coasts in large numbers. Butterflies reaching our Southern Counties early in the year, will lay eggs to produce a home brood in August. Adults from the Mediterranean continue to arrive until late autumn. Usually seen in ones or twos, it has a direct fast flight and rarely settles for long. The upperside is only seen when flying, as it's wings always remain closed when at rest. More likely to be encountered in the South, but individuals will venture Northward to all parts of the country.

Flight Season :   April - November

Caterpillar's Foodplant :   Clover, Vetches, Lucerne.

CLOUDED YELLOW

MIDSUMMER MIGRANT, THE CLOUDED YELLOW

BRIMSTONE

EARLY SPRING FLIGHT, THE BRIMSTONE

Bright yellow with a unique shape, this is the original 'butter' - fly. An unmistakeable sight in early spring when hibernating adults wake to the first warm day of the year. Absent from Scotland and most of Northern England, the brimstone is limited by the distribution of it's foodplant, buckthorn. The eggs are laid in Spring and the first of the summer adults emerge in late July. These butterflies will hibernate through the winter and can still be seen on the wing, the following June.

Flight Season : February to November

Caterpillar's Foodplant : Buckthorn

BRIMSTONE

FRESHLY EMERGED BRIMSTONE - JULY

HAIRSTREAKS

LYCAENIDAE

Named after the fine white lines which cross their underside the Hairstreaks are small butterflies, living around trees and bushes. Apart from the Green Hairstreak, they have "tails" on the hind wing and are usually seen flying high, rarely venturing down to ground level.

PURPLE HAIRSTEAK	*NEOZEPHYRS QUERCUS*
BROWN HAIRSTREAK	*THECLA BETULAE*
WHITE-LETTER HAIRSTREAK	*SATYRIUM WALBUM*
BLACK HAIRSTREAK	*SATYRIUM PRUNI*
GREEN HAIRSTEAK	*CALLOPHRYS RUBI*

PURPLE HAIRSTREAK

PURPLE HAIRSTREAKS, MALE AND UNDERSIDE

In July and August this small butterfly can be seen circling around the tops of oak trees and will fly till sunset on warm evenings. I've sometimes seen it in large numbers feeding from the sap of young Acorn cups, but only occasionally coming down to eye level. The female, shown on the opposite page, has a bright patch of purple at the base of the forewings. The male is dark dusky brown with a purple sheen. This is the most widespread of the hairstreaks and is common in oak woods throughout its range. Found in all Southern counties, South Midlands, Wales and East Anglia. Elsewhere, very local.

Main Flight Season : July and August

Caterpillar's Foodplant : Oak buds and leaves

PURPLE HAIRSTREAK

PURPLE HAIRSTREAK - THE OAK CANOPY

BROWN HAIRSTREAK

NECTARING ON HEMP AGRIMONY

This is a rare butterfly which frequents thickets of Blackthorn. Adults like to congregate around a larger tree, often Ash, within an overgrown hedgerow. The Brown Hairstreak flies throughout August and September, but doesn't often take to the wing, so is easily overlooked. Despite it's drab name it has a beautiful golden orange underside. Its colonies are very localised, in Southern England and West Wales.

Flight Season : August - September

Caterpillar's Foodplant : Blackthorn

BROWN HAIRSTREAK

THE BROWN HAIRSTREAK IN WILTSHIRE

WHITE-LETTER HAIRSTREAK

FIELD SKETCH - WHITE-LETTER HAIRSTREAK

This attractive Hairstreak is named after the white letter 'W' on the hindwing of the underside. Its lifecycle is dependent upon Elms, so it has suffered a rapid decline in numbers since the spread of Dutch Elm Disease. Elusive and never common, but found in scattered colonies over much of Southern England, the Midlands and Wales. Usually seen flying round the tops of young elms, but sometimes flies down to nectar on bramble or thistle beside the trees. In the future, it's population may well be increased by the introduction of disease resistant species of Elms.

Flight Season : July - August

Caterpillar's Foodplant : Elm flowers and leaves

WHITE-LETTER HAIRSTREAK

THE ELM HEDGEROW

BLACK HAIRSTREAK

NECTARING ON WILD PRIVET

Very rare, found only in a small band of colonies, north east of Oxford. Most of these are managed nature reserves where it survives on tall Blackthorn hedgerows and along sheltered woodland edges. The underside is very similar to the White Letter Hairstreak, but it has a more rounded wing shape and a line of black spots within the orange border. The Black Hairstreak is usually seen flying around the tops of old Blackthorn bushes, but will sometimes fly down to nectar on wild privet. It's wings always remain closed when settled.

Flight Season : mid June - mid July

Caterpillar's Foodplant : Blackthorn

BLACK HAIRSTREAK

THE BLACK HAIRSTREAK IN OXFORDSHIRE

GREEN HAIRSTREAK

PERFECT CAMOUFLAGE - THE GREEN HAIRSTREAK UNDERSIDE

This is the smallest of the hairstreaks and has an unmistakeable bright green underside. The wings are always kept closed so the butterfly is well camouflaged in any leafy situation. It has a fast jerky flight, and appears dusky brown and difficult to follow. Commonest in the southern counties and very local elsewhere, although large colonies exist in West Scotland. This is the only hairstreak to be found frequently at ground level. Its usual habitats are gorse covered valleys, sheltered bushy heathland, and along woodland edges.

Flight Season : May - June

Caterpillar's Foodplant : Gorse, Rockrose, Bird's Foot Trefoil, Dogwood, Buckthorn

GREEN HAIRSTREAK

THE GREEN HAIRSTREAK IN MAY

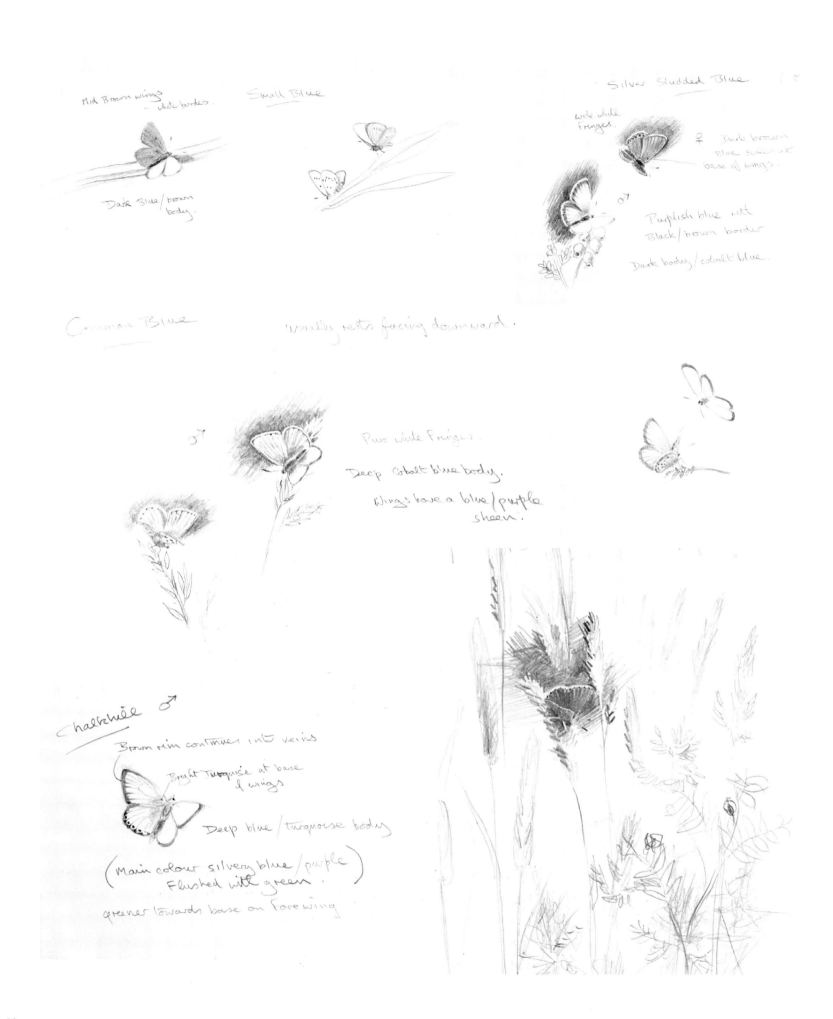

THE COPPERS

LYCAENIDAE

Britain has only one species in this family, a diminutive butterfly easily recognised by it's brilliant metallic orange wings.

SMALL COPPER *LYCAENA PHLAEAS*

THE BLUES

LYCAENIDAE

Mainly found on wild grassland where they live in compact colonies, rarely straying from their breeding area. They are very active, flying fast and low to the ground. The exception is the Holly Blue which can be seen fluttering around trees and bushes, often in gardens. The striking metallic blue of some species is in strong contrast to the brown females. However all species and sexes have a linking feature, their beautiful, intricate undersides, marked with rings and spots.

SILVER-STUDDED BLUE *PLEBEIUS ARGUS*
BROWN ARGUS *ARICIA AGESTIS*
NORTHERN BROWN ARGUS *ARICIA ARTAXERXES*
COMMON BLUE *POLYOMMATUS ICARUS*
ADONIS BLUE *POLYOMMATUS BELLARGUS*
CHALKHILL BLUE *POLYOMMATUS CORIDON*
HOLLY BLUE *CELASTRINA ARGIOLUS*
SMALL BLUE *CUPIDO MINIMUS*
LARGE BLUE *MACULINEA ARION*

THE DUKE OF BURGUNDY FRITILLARY

HAMEARIS

Not a true Fritillary, this small butterfly is a link between Lycaenidae and Nymphalidae. In size and life cycle it resembles the Coppers and Blues, but it has only four walking legs and fritillary-like wing markings.

DUKE OF BURGUNDY FRITILLARY *HAMEARIS LUCINA*

SMALL COPPER

SMALL COPPER UNDERSIDE

A small species, but conspicuous due to it's bright orange forewings resembling shining copper. Common and widespread as far North as Yorkshire but more local in Scotland and Ireland. Found in almost any open habitat where Sorrel is growing. I've always seen this butterfly in ones or twos, and only rarely in large numbers. It has a territorial nature, darting off from it's perch after any small butterfly that passes. Only to return to the same spot moments later.

Flight Seasons : 2 or 3 broods May, July - August, September - October

Caterpillar's Foodplant : Common Sorrel and Sheep's Sorrel

SMALL COPPER

READY TO FLY, THE SMALL COPPER

SILVER-STUDDED BLUE

NEW FOREST HEATHLAND, THE SILVER-STUDDED BLUE

The male is silvery blue with black borders and pure white fringes and is easily distinguished from the other blues by it's smaller size. The female is dusky brown with a margin of orange spots, similar to the Brown Argus, but less well defined. This is primarily a heathland species, preferring damp ground, where it sometimes occurs in vast numbers and is often the only species present. On warm evenings I've seen hundreds perched on heather with wings outstretched catching the last rays of the sun, a truly memorable sight!

The main stronghold for this localised species is on the heathlands of Dorset, Hampshire, Sussex and Surrey. It also occurs in a few isolated colonies on coastal chalk and limestone, as far North as Anglesey in the west and Norfolk in the east.

Flight Season : July - August

Caterpillar's Foodplant : Mainly Heather, Gorse and Bird's Foot Trefoil

SILVER-STUDDED BLUE

SILVER-STUDDED BLUES ON BELL HEATHER

BROWN ARGUS

A WARM SUMMER EVENING, BROWN ARGUS AND UNDERSIDE

A local butterfly restricted mainly to Southern England and a few localities around the Welsh Coast. Both male and female have pure brown wings with an edging of clear orange spots. The Brown Argus is very small and difficult to follow when in flight. Often seen basking in the sun with wings outstretched. It occurs in many open habitats with sparse low growing vegetation, and an abundance of Rockrose, but is most frequent on steep south facing slopes. Less commonly this species is found along field edges and woodland clearings where the caterpillar feeds on Crane's Bill.

Flight Season : May - June second brood August - September.

Caterpillar's Foodplant : Common Rock-rose, less frequently - Stork's-bill , Crane's-bill

BROWN ARGUS

JUST RESTING, THE BROWN ARGUS

NORTHERN BROWN ARGUS

ROOSTING IN THE HOLLOW, THE NORTHERN BROWN ARGUS

Once known as the Scotch White Spot, this is very similar to the Brown Argus. The obvious difference between these two species is a white discal spot on the upper wing. Also the white spot on the underside are usually without black pupils. It inhabits hillsides and lower mountain slopes where scree and tight grazing encourage the growth of Rock-rose. Prefers a south facing aspect. It's main stronghold is in Central and Eastern Scotland. Smaller scattered colonies can also be found in Southern Border Counties and in a few localities in the North of England.

Flight Season :   mid June - July

Caterpillar's Foodplant :   Common Rock-rose

NORTHERN BROWN ARGUS

NORTHERN BROWN ARGUS ON ROCKROSE

COMMON BLUE

BASKING IN THE SUN, THE COMMON BLUE

This is the most common of our native blues and is found throughout Britain in a variety of open habitats. It's presence in any area is closely related to an abundance of it's main foodplant, Bird's foot trefoil. The male is bright mid-blue with a hint of violet. The female is generally brown, often flushed with blue, to a variable extent and has a margin of orange spots. The intricately marked underside has a margin of orange spots in both sexes. (see Evening Shadows, Martin Down page 7) An occasional visitor to gardens as this butterfly tends to wander from its wild breeding sites.

Flight Season : South : Two broods May - June, August September
North and Ireland : One Brood July - August

Caterpillar's Foodplant : Bird's Foot Trefoil also on Rest Harrow and Black Medick

COMMON BLUE

COMMON BLUES, MALE AND FEMALE

ADONIS BLUE

FEASTING ON MARJORAM, THE ADONIS BLUE

Once seen never forgotten, the wings of the male are an intense metallic turquoise which rarely occurs in nature. The female is chocolate brown with an edging of orange spots. In both sexes dark veins run through the white outer fringes of the wings. The markings of the underside are very similar to those of the Common Blue with which it can be confused. Unfortunately a rare and very localised species confined to a few chalk or limestone sites in the mid south and south east. This butterfly is one of my favourite sights. It is usually encountered on warm, south facing slopes or hollows lightly grazed by sheep or rabbits.

Flight Seasons :   Mid May - June   Second Brood - Mid August - September

Caterpillar's Foodplant : Horseshoe vetch

ADONIS BLUE

DOWNLAND GEM, THE ADONIS BLUE

CHALKHILL BLUE

CHALKHILL BLUES, A MATING PAIR

As it's name implies, this species is found on chalk downland and is confined mainly to the counties of the mid south and South East. The wings of the male are a beautiful pale silvery blue, with a hint of turquoise. The female is a rich brown with an indistinct edging of orange spots. Larger than the Common Blue, it enjoys flower rich, south facing slopes. In some years the butterflies emerge in vast numbers. I've seen them in thousands on many Hampshire sites.

Flight Season :   July - Mid September

Caterpillar's Foodplant :   Horse Shoe Vetch

CHALKHILL BLUE

ON A DORSET HILLSIDE

HOLLY BLUE

FIELD NOTES - HOLLY BLUES

Common and Widespread in the Southern half of Britain. Otherwise, local, extending up to Southern Scotland. Mainly seen flying among hedgerows and bushes from open woodland to town gardens. In my garden, these silvery-blue butterflies flutter around the ivy growing up the side of the house. They usually settle on foliage with wings closed, but sometimes, in warm sunshine, partially open them to reveal the attractive violet-blue upper side. The forewings of the male have a narrow black edge, whereas those of the female have a wide black border. The exquisite silvery underside is marked with a sprinkling of tiny black dots.

Flight Season : First brood - April - June   Second brood - July - August

Caterpillars Foodplant : Mainly holly and ivy. Occasionally on the buds of other shrubs.

HOLLY BLUE

ALONG THE HEDGEROW - HOLLY BLUE PAIR

This species is often associated with Holly and Ivy. In Spring the eggs are usually laid on the flower heads of Holly and later, the summer brood will lay on the compact buds of Ivy.

SMALL BLUE

FIELD SKETCH, SMALL BLUES

    This is Britain's smallest butterfly. Both male and female upper wings are plain brown with only a hint of blue. The distinctive underwing colour is pale silvery grey. I usually see them perched on bushes or taller grasses in warm sheltered hollows. The Small Blue's main stronghold and it's largest colonies are on the chalk and limestone hills of the mid south. Elsewhere it is extremely local and restricted mainly to a few coastal sites where Kidney vetch is abundant.

    A few sites occur along the east coast of Scotland but is absent from the west. Colonies can often be very small, supporting less than thirty individuals.

Flight Season :   Mid May - June August

Caterpillar's Foodplant :   Kidney Vetch

SMALL BLUE

SMALL BLUES AT MARTIN DOWN

LARGE BLUE

LARGE BLUE, UNDERSIDE

This butterfly declined rapidly in numbers and was considered extinct by 1980. However it has now been reintroduced to some suitable sites and is being carefully monitored to check it's progress. Hilary and I were lucky enough to see these magnificent blues at one of the new colonies in Somerset. Wing colour in both male and female is a deep steely blue and this species is much larger than it's close relatives. The caterpillar begins feeding on Wild Thyme, but is soon discovered and carried away by a particular species of red ant, Myrmica Sabuleti. It is taken down into the ant nest, where it is nurtured and allowed to feed on ant grubs. After spending the winter as a chrysalis underground, the adult finally emerges in June.

Flight Season : mid June - July

Caterpillar's Foodplant : Wild thyme, but also relies on an association with ants.

LARGE BLUE

NECTARING ON WILD THYME, SOMERSET

DUKE OF BURGUNDY FRITILLARY

COWSLIP VISITOR, THE DUKE OF BURGUNDY FRITILLARY

Not a true 'Fritillary', the Duke of Burgundy is Britain's only member of the 'Metalmarks', and is more closely related to the Blues and Coppers. Despite having Fritillary-like markings it is easily identified by it's small size. Once far more widespread, it is now virtually confined to the mid South, where it lives in small discreet colonies in areas where Cowslip or Primrose are abundant. It's favourite haunts are bushy sheltered areas, and sunny clearings in woodland. Likes to bask in the sun and will dart off after any butterfly which passes.

Flight Season : May - June

Caterpillar's Foodplant : Cowslip and Primrose

DUKE OF BURGUNDY FRITILLARY

DUKE OF BURGUNDY FRITILLARIES

EMPERORS, WHITE ADMIRALS AND VANESSIDS

NYMPHALIDAE

The Nymphalids are known as the "brush footed butterflies". Their front pair of legs are reduced to short sensory brushes which they use for taste, and in the case of females, to detect the correct foodplant. They have, in effect, only four walking legs.

The Purple Emperor is a large woodland butterfly, famous for the male's iridescent purple sheen. It spends most of its life high in the tree canopy.

There are numerous species of White Admirals worldwide, but Britain has just the one. It's dark brown wings have a pure white band running through the upper and lower wings. It is a woodland butterfly which has spread to many new locations over the last century.

The Vanessids are known for their bright orange and red colours, though their underwings are drab and well camouflaged. This is particularly noticeable in the Comma, the Peacock and the Small Tortoiseshell, which hibernate so successfully. They are robust butterflies with a strong flight, frequently interspersed by gliding. The Vanessids do not live within the confines of a colony, but wander and migrate to wherever habitats are suitable. All are nectar loving species, frequently found in gardens. They also enjoy over ripe fruit and are regular visitors to orchards. There is no discernable difference in wing pattern between the male and female, in our British species.

PURPLE EMPEROR	*APATURA IRIS*
WHITE ADMIRAL	*LIMENITIS CAMILLA*
PAINTED LADY	*VANESSA CARDUI*
RED ADMIRAL	*VANESSA ATALANTA*
SMALL TORTOISESHELL	*AGLAIS URTICAE*
PEACOCK	*INACHIS IO*
COMMA	*POLGYONIA C-ALBUM*

PURPLE EMPEROR

GLIDING THROUGH THE SALLOWS, THE FEMALE EMPEROR

This magnificent Butterfly needs large tracts of open deciduous woodland where tall Sallows grow along sunny rides and glades. It is a rare and local species, confined almost exclusively to central Southern England. The adult males gather in the tree canopy to defend a chosen perch on a favourite tree, often at the highest point in the wood. The female can be seen gliding through the sallow branches, landing momentarily to lay eggs on the leaves. Collectors used to travel great distances with nets on 30 foot extendable poles to catch this great prize. Males can sometimes be tempted down to the ground where they feed on decaying matter or mineral salts on a pathway.

Flight Season : July - August

Caterpillar's Foodplant : Sallow

PURPLE EMPEROR

HIGH IN THE TREETOPS

WHITE ADMIRAL

HONEYSUCKLE, HAZEL AND WHITE ADMIRALS

A handsome butterfly with a distinctive gliding flight, not dissimilar to the Purple Emperor, but much smaller. It looks exceptional when freshly emerged, a dark velvet brown with a contrasting pure white band. Found in the Southern half of England, being less frequent in the East and extreme West. The White Admiral frequents sunny glades and woodland edges, where honeysuckle rambles through the undergrowth. Can often be seen flying into shady areas when egg laying. This is one of the few butterflies which has greatly expanded its range over the last century.

Flight Season : July - mid. August

Caterpillar's Foodplant : Honeysuckle

WHITE ADMIRAL

AT THE FOREST EDGE, WHITE ADMIRAL

PAINTED LADY

VISITING THE SPEAR THISTLES

One of our commonest migrants, the Painted Lady can be seen in almost any habitat throughout the British Isles. It loves to bask in the sun with wings outstretched and is a common visitor to garden flowers, especially Buddleia. This species arrives early in the year in small numbers. These adults will lay eggs to produce a second brood of British raised ladies, emerging from their chrysalids in mid summer. They can be seen until late Autumn, but cannot survive the British winter.

Flight Season :	April - June Second brood August - October

Caterpillar's Foodplant :	Mainly Thistle.

PAINTED LADY

THE PAINTED LADY IN DORSET

RED ADMIRAL

RED ADMIRALS ON THISTLES

The pure red band against a velvet black background makes this species unmistakeable. One of our largest butterflies, which makes it's first appearance in April and May. These individuals are migrants from the continent which will lay eggs to produce a summer emergence in July and August. Adults continue to arrive from abroad so numbers build up as the year progresses. Found in all habitats throughout the country, but most frequently noted in gardens. They are particularily fond of Buddleia and Michaelmas Daisies. I sometimes see this butterfly as late as Early December and again in February, which suggests that a few may survive the winter.

Flight Season : April -November

Caterpillar's Foodplant : Stinging Nettle

RED ADMIRAL

A FLASH OF RED

SMALL TORTOISESHELL

WILD CORNER WITH SMALL TORTOISESHELL

This is one of Britain's commonest and most well known butterflies. Its vivid colours make it an attractive sight, especially in early spring after hibernation. The bright upperwings are in contrast to its subdued and well camouflaged underside (see garden visitors page 196). In winter, the sleeping butterflies are frequently seen in houses and outbuildings. Although found in all habitats, the Small Tortoiseshell is especially noted in gardens, where it nectars in large numbers on Buddleia and later on Michaelmas Daisies and Sedum. Its black and yellow spiny caterpillars are a common sight on nettle patches.

Flight Season : March - October

Caterpillar's Foodplant : Stinging Nettle

SMALL TORTOISESHELL

NECTARING ON VERBENA, THE SMALL TORTOISESHELL

PEACOCK

EARLY SPRING, THE PEACOCK

The 'eyes' on this large robust butterfly distinguish it from any other British species. It also has a unique velvety black underside. (see garden visitors, page 198) The Peacock will often remain active until the first frosts of Winter and is one of the earliest butterflies to be seen in Spring after hibernation. Found in most habitats throughout the country, but absent from the North of Scotland. This handsome butterfly loves to visit gardens, where it is most commonly seen on Buddleia. The large black spiny caterpillars are gregarious and are often encountered in webs on stinging nettles.

Flight Season :   March - May   July - November

Caterpillar's Foodplant : Stinging nettle

PEACOCK

PEACOCK ON HEMP AGRIMONY

COMMA

AUTUMN CAMOUFLAGE, THE COMMA

The Comma is one of Britain's hibernating butterflies and is well camouflaged by it's broken outline and it's resemblance to a dead leaf. It has a dark marked orange - brown upperside and a distinct white "comma" mark on the hindwing of the underside. Can be mistaken for a fritillary when in flight. This butterfly has made a remarkable recovery after it's slow decline to virtual extinction by 1910. Since then it has spread to all southern and midland counties and throughout Wales. Commonly found in sunny wooded areas, copses, hedgerows and gardens. Loves to visit orchards for fallen fruit, and also enjoys ripe blackberries and nectaring on garden flowers.

Flight Season : March - October

Caterpillar's Foodplant : Stinging Nettle, Hops and Elm.

COMMA

THE COMMA BUTTERFLY

FRITILLARIES

NYMPHALIDAE

The Nymphalids are known as the "brush footed butterflies". Their front pair of legs are reduced to short sensory brushes which they use for taste, and in the case of females, to detect the correct foodplant. They have, in effect, only four walking legs.

The Fritillaries are recognised by their bright orange or golden brown uppersides, with an intricate patterning of dark spots and bands. The undersides are also intricately marked, some having silver spots or washes. They keep within their compact breeding colonies and are very rarely seen in gardens. On the wing, they are generally fast, with frequent glides.

PEARL-BORDERED FRITILLARY	*BOLORIA EUPHROSYNE*
SMALL PEARL-BORDERED FRITILLARY	*BOLORIA SELENE*
DARK GREEN FRITILLARY	*ARGYNNIS AGLAIA*
HIGH BROWN FRITILLARY	*ARGYNNIS ADIPPE*
SILVER-WASHED FRITILLARY	*ARGYNNIS PAPHIA*
MARSH FRITILLARY	*EUPHYDRAS AURINIA*
HEATH FRITILLARY	*MELITAEA ATHALIA*
GLANVILLE FRITILLARY	*MELITAEA CINXIA*

PEARL-BORDERED FRITILLARY

VISITING THE BUGLE FLOWERS

UNDERSIDE, WITH CENTRAL SILVER MARK

Primarily a woodland butterfly, preferring sunny rides and newly cleared areas where violets can flourish. Previously more common, when coppicing was a routine practice. Local and confined to scattered colonies in the western half of Britain and the extreme south. The main distinguishing feature between this and the Small Pearl-bordered is the underside pattern. Both species have a border of seven 'pearls', but this species has one large silver mark on the middle of the wing, instead of the many found on the Small Pearl-bordered. Recently it has made a dramatic recovery in parts of the New Forest, where I've seen fifty or more in newly cleared plantation blocks. It has a fast low flight, flitting and Gliding until it comes to rest on a suitable Bugle flower or a sunny spot.

Flight Season : May - June

Caterpillars Foodplants : Common Dog Violet and Marsh Violet

PEARL-BORDERED FRITILLARY

WOODLAND BUTTERFLY, THE PEARL-BORDERED FRITILLARY

SMALL PEARL-BORDERED FRITILLARY

ABOVE THE LOCH, THE SMALL PEARL-BORDERED FRITILLARY

Contrary to it's name, the Small Pearl-Bordered Fritillary is about the same size as the Pearl-bordered. The upperside markings are also very similar, but this species has a complex pattern of Silver Spots on the underside, in addition to its border of silver 'pearls'. This is a very local butterfly confined mainly to the damper western parts of Britain, but not found in Ireland. Shares the same habitats and often flies alongside the Pearl-Bordered, but can also be found in wild open areas and on coastal cliffs. In West Scotland I've seen it on open moor land, flying in sheltered bracken lined gulleys and damp valleys.

Flight Season :   June - mid July

Caterpillars Foodplants :   Common Dog Violet and Marsh Violet

SMALL PEARL-BORDERED FRITILLARY

SMALL PEARL-BORDERED FRITILLARIES

DARK GREEN FRITILLARY

UNDERSIDE, SHOWING THE SILVER SPOTS

FIELD SKETCH - FRITILLARIES ON THE DOWNS

This medium sized Fritillary is named after its attractive greenish underside which is adorned with clear silver spots. Occurs throughout Britain, particularly in coastal locations, but is virtually absent from the East of England. It is usually seen flying fast over open downland or grassland in July and August. Found also, but less frequently, on moorland and along sunny woodland rides. The male is a bright orange brown, the female is larger and usually less bright. They nectar frequently on knapweed and scabious, where they can be observed at close quarters.

Flight Season : July - August

Caterpillars Foodplants : Hairy, Common Dog and Marsh Violets

DARK GREEN FRITILLARY

DARK GREEN FRITILLARY ON DWARF THISTLE

HIGH BROWN FRITILLARY

FLIGHT OF THE HIGH BROWNS, THE MALVERN HILLS

This fritillary is easily mistaken for the Dark Green, but the underside has a browner suffusion and a band of ringed spots next to the outer edge of silver spots. It used to be more widespread, but it is now an extremely local species, confined to western England and Wales. It has a very fast flight and is usually seen skimming over bracken covered hillsides. In order to breed successfully the caterpillar needs large tracts of sheep grazed or cut bracken. Over wintering as an egg , it needs a sheltered spot near a violet bed, where the caterpillar can feed in the spring. Also survives in a few wooded river valleys, frequenting open sunny glades where it likes to nectar on thistles or bramble.

Flight Season : July

Caterpillars Foodplants : Common Dog Violet, Hairy Violet

HIGH BROWN FRITILLARY

BRACKEN HILLSIDE WITH HIGH BROWN FRITILLARIES

SILVER-WASHED FRITILLARY

SUNLIGHT THROUGH THE WINGS   ON A NEW FOREST TRACK, THE MALE SILVER-WASHED FRITILLARY

This is the largest of our Fritillaries, usually seen nectaring on thistle or bramble along the edges of sunny forest glades. Often flies high into the tree canopy. Widely distributed throughout the mid south and south west of England, Wales and Ireland. Prefers large tracts of woodland where, in warm years, it can occur in very large numbers. It has an unusual courtship flight. As the female flies fast and purposefully, the male circles round her in tight vertical loops. Named after the streaks of silver running through it's greenish under wing, quite distinct and different from the clear silver spots of our other large fritillaries. (see Mating Pair, Silver-Washed Fritillaries, page 32) Hibernates as a caterpillar, on a mossy tree trunk near violets.

Flight Season :   July - August

Caterpillars Foodplants :   Common Dog Violet

SILVER-WASHED FRITILLARY

WOODLAND CLEARING, SILVER-WASHED FRITILLARIES

MARSH FRITILLARY

MARSH FRITILLARIES IN DAMP MEADOW HABITAT

This attractive and intricately marked fritillary has suffered a rapid decline in recent years. It occurs where Devil's Bit Scabious is abundant, in two contrasting habitats, damp fields and dry chalk downland. Found only in the western half of Britain, chiefly Devon, Cornwall and South Wales. Elsewhere it survives in a few isolated colonies. This species is not as active as the other fritillaries, the female frequently sitting still for long periods.

Flight Season : mid May - July

Caterpillars Foodplants : Devil's Bit Scabious

MARSH FRITILLARY

MARSH FRITILLARIES

HEATH FRITILLARY

HEATH FRITILLARIES IN COPPICED WOODLAND

One of Britain's rarest butterflies, with a colony in Kent, one in Essex and a handful in Devon and Cornwall. It's usual habitat is in sunny sheltered woodland, where Cow Wheat is abundant. It likes recently felled clearings and once was more widespread, but slowly disappeared with the decline in coppicing. On Exmoor it survives however on open hillsides, in a few sheltered coombes. Browner than other Fritillaries and frequently glides when in flight.

Flight Season : June and July

Caterpillars Foodplants : Mainly Cow Wheat. Occasionally on Ribwort Plantain or Foxglove.

HEATH FRITILLARY

THE HEATH FRITILLARY ON EXMOOR

GLANVILLE FRITILLARY

NECTARING ON THRIFT

Apart from occasional stray colonies on the Hampshire coast, this butterfly is confined exclusively to the Isle of Wight. Here it inhabits rough ground on the warm undercliff of the Southern coast. Colonies may contain several hundred individuals. The upper wings are orange brown banded with dark markings, not dissimilar to the Heath Fritillary. The beautiful and unmistakable underside is white, crossed by two bands of orange brown. The Glanville's flight is fast and low to the ground, gliding frequently.

Flight Season : mid May - June

Caterpillars Foodplants : Ribwort Plantain

GLANVILLE FRITILLARY

GLANVILLE FRITILLARY ON THE UNDERCLIFF, ISLE OF WIGHT

THE BROWNS

SATYRIDAE

Apart from the Marbled White, these butterflies are, as would be expected, brown. Closely related to the Nymphalids, they have only four walking legs. Members of this family are easily identified by their "eyespots" - black dots with a white pupil. These "eyespots" are usually set in a patch or ring of yellow or orange, forming a row near the wing edges. In the case of the Meadow Brown, Gatekeeper and Small Heath they are reduced to a single "eye" on the forewing. The browns have a characteristic slow flight, appearing to hang in the air, as they flap lazily over the tops of the grasses.

SPECKLED WOOD	*PARARGE AEGERIA*
MEADOW BROWN	*MANIOLA JURTINA*
GATEKEEPER	*PYRONIA TITHONUS*
MARBLED WHITE	*MELANARGIA GALATHEA*
RINGLET	*APHANTOPUS HYPERANTUS*
SMALL HEATH	*COENONYMPHA PAMPHILUS*
LARGE HEATH	*COENONYMPHA TULLIA*
GRAYLING	*HIPPARCHIA SEMELE*
WALL	*LASIOMMATA MEGERA*
SCOTCH ARGUS	*EREBIA AETHIOPS*
MOUNTAIN RINGLET	*EREBIA EPIPHRON*

SPECKLED WOOD

THE WOODLAND FLOOR

Common and often abundant where it occurs, the Speckled Wood is a shade loving species. It prefers leafy places in woods and lanes, where it likes to bask in patches of dappled sunlight. This butterfly has a dancing flight and is the only species to be found on heavily shaded woodland pathways. It sometimes ventures into more open areas and gardens, seeking out any shady corner. Its stronghold is in Southern England and the Midlands, though it is more localised in the Eastern counties. The Speckled Wood has recently been expanding its range and now has a thriving population in West Scotland and Ireland

Flight Season : April - September with three broods.

Caterpillar's Foodplant : Grasses especially Common Couch grass

SPECKLED WOOD

THE SPECKLED WOOD IN FOREST GLADE

MEADOW BROWN

THE MEADOW BROWN, MALE AND UNDERSIDE

This is the most common of the 'browns' and is found throughout Britain in any sunny area where wild grasses are present. The male is the darker and slightly smaller of the sexes. The female is more colourful with a larger patch of pale orange on the forewing below the white pupilled black eyespot. The Meadow Brown has a lazy flight and is often encountered in large numbers in hayfields and on wild grassland. In weak sunshine and on bright evenings this species will bask with wings open, but normally it rests with closed wings. Favourite nectar sources include knapweed, thistles, marjoram, scabious, and hawkweed.

Flight Season :   Mid June - mid September

Caterpillar's Foodplant :   a variety of wild grasses

MEADOW BROWN

NECTAR SIPPING - MEADOW BROWNS

GATEKEEPER

THE LAST THISTLE FLOWER

The Gatekeeper, also known as the Hedge Brown, is often seen flying in company with the Meadow Brown, but is less fond of open areas, preferring some shelter from hedgerows and bushes. However it is found in open heathy places, where it settles along the bracken margins. Being an abundant species, it often strays into gardens in country areas. It is a brighter orange colour and slightly smaller than the Meadow Brown. Common and widespread in the south but in northern counties, more local and confined to the western coast. Absent from Scotland.

Flight Season : July to September

Caterpillar's Foodplant : Meadow grasses, Bents and Fescues

GATEKEEPER

GATEKEEPERS IN THE NEW FOREST

MARBLED WHITE

GRASSLAND PAIR

The Marbled White is the unrepresentative member of the "Browns" family, having a black and white chequered pattern. However, it has the "Browns" distinctive flapping flight and stays within the boundaries of a colony. It is a butterfly of high summer, usually associated with open areas of wild grassland, being particularly plentiful on chalk and limestone soils. A superb sight when seen nectaring or basking in groups, with open wings. I've sometimes seen ten or more on single thistle patch. This is a Southern species, its stronghold being the Central and South West counties, with a few scattered colonies elsewhere. Absent from Scotland and Ireland.

Flight Season : mid. June - August

Caterpillar's Foodplant : wild grasses - Red Fescue, Sheep's Fescue, and Tor Grass

MARBLED WHITE

KNAPWEED AND MARBLED WHITES

RINGLET

RINGLETS IN A SHADY WOODLAND CLEARING

The Ringlet appears dark brown on the wing and is easily mistaken for the similar sized Meadow Brown. In both these species old worn specimens at the end of the season are quite pale, whereas freshly emerged individuals can be very dark. When settled the Ringlet invariably rests with it's wings closed and is then easily identified by it's attractive row of ringed eyespots. Likes dappled shade where wild grasses are growing lush and tall. Can become very numerous in damp areas. Locally common in Southern and Eastern England, Wales and Ireland. Rare in the North of England, but more common in Southern Scotland.

Flight Season : July - mid August

Caterpillar's Foodplant : A wide variety of grasses including Cock's Foot, Annual Meadow Grass and Common Couch.

RINGLET

TWO RINGLETS

SMALL HEATH

THE DORSET COASTLINE, SMALL HEATH

The smallest of the Browns preferring downland, heathland and rough grassy areas. Easily recognised, being Britain's only small pale brown butterfly. Common and widespread throughout the country. When disturbed it flits close to the ground and quite suddenly drops to a new resting place, always with wings closed. I've seen the Small Heath in greatest numbers on coastal hillsides and extensive moorland.

Flight Season : May - July   August - October   two broods

Caterpillar's Foodplant : Various wild grasses, Bents and Fescues

SMALL HEATH

SMALL HEATH ON HAMPTON RIDGE

LARGE HEATH

LARGE HEATH - THE SCOTTISH RACE

This butterfly was once known as the Marsh ringlet. In flight, it looks drab and considerably larger than the Small Heath. Scotland is the large heath's main stronghold where it inhabits mires, damp moorland, and mountain slopes. Further south and in Ireland, scattered colonies exist where peat bogs remain intact. Shropshire is England's most southerly range of this species, and Welsh colonies live in Snowdonia. Southern specimens have the larger, clearer eyespots.

Main Flight Season : Mid June - July

Caterpillar's Foodplant : Hare's-tail Cottongrass.

LARGE HEATH

THE LARGE HEATH IN SHROPSHIRE

GRAYLING

COURTSHIP ON THE HEATH

This is the largest of the browns and always settles with wings closed, usually on a patch of bare ground. When disturbed flies rapidly, but soon drops down and disappears. It's mottled underside and it's habit of leaning towards the sun gives a good camouflage. Occasionally I've been lucky to catch a glimpse of the Grayling's upper wings during courtship displays. Common around most of the British coast on rocky areas and sand dunes, also in large numbers on southern heathland.

Flight Season : July to September

Caterpillar's Foodplant : Wild grasses, Bristle Bent, Marram, Red and Sheep's Fescue

GRAYLING

HEATHLAND BUTTERFLY, THE GRAYLING

WALL

INTRICATE UNDERSIDE, THE WALL BROWN

This species is frequently seen basking on walls or any other suitable sun-baked location. Its habitats include rough open grassland, wasteland, quarries and eroded cliffs. Also known as "The Wall Brown", it is the brightest member of this family, and is sometimes mistaken for a fritillary. It often settles and basks on pathways where it is easily identified by it's eye spots. Well distributed in England, Wales and Ireland, but rarer in the North where it occurs only in Seaboard Counties.

Flight Season :　　　　　　May - June　Mid July - September

Caterpillar's Foodplant :　　Various Coarse grasses

WALL

COASTAL HABITAT, THE WALL BUTTERFLY

SCOTCH ARGUS

RESTING POSITION, THE SCOTCH ARGUS

As it's name suggests this medium sized butterfly is a true Northern Species. It is locally common in Southwest and Central Scotland and is widespread throughout the Western Highlands and Islands. It's usual haunts are damp grassy valleys and lower mountain slopes, preferring the sheltered edges of plantations. Even when large numbers are flying, this species has a habit of disappearing quite suddenly with a passing cloud shadow and reappearing just as suddenly when the sun re-emerges.

Flight Season : Mid. July - August

Caterpillar's Foodplant : Purple Moor Grass, Blue Moor Grass

SCOTCH ARGUS

ON A SHELTERED NORTHERN HILLSIDE, THE SCOTCH ARGUS

MOUNTAIN RINGLET

RESTING ON MAT GRASS, MOUNTAIN RINGLETS

A small Northern species found only at high altitudes, up to 1000 metres. Extremely local being confined to a few localities in the central Highlands and the Lake District. It lives on exposed mountain sides and prefers damp south facing hollows and gullies. Flies only in bright sunshine, a habit which makes it difficult to locate in such cloudy areas! Has a slow flight keeping close to the ground.

Flight Season :   Mid June - July

Caterpillar's Foodplant :   Mat Grass

MOUNTAIN RINGLET

RESIDENT OF THE SCOTTISH HIGHLANDS

Oil 36 × 40 in

THE COMMON BLUE BUTTERFLY

FIRST PRIZE WINNER OF NATIONAL COMPETITON, "ART IN NATURE" 1996

SMALL BUTTERFLIES IN A LARGE LANDSCAPE

SMALL COPPERS　　　　　　　　　　Oil　　16 × 20 in

Oil 30 × 30 in

THE MARBLED WHITE

THE SECRET TRACK　　　　Oil　36 × 36 in

In damp riverside fields, a sea of yellow buttercups make their annual display in May and June. This coincides with the appearance of one of our brightest spring butterflies, the Orange Tip. In search of a mate, the unmistakable male flies leisurely along the course of the river.

THE MARBLED WHITE
Page 166
In this painting, the verge of an ancient track above Barford St. Martin, supports a wealth of wild plants along it's length. Often running between cultivated fields, it creates a natural oasis, home to a large range of common butterflies. Here, the black and white chequered Marbled White, takes a rest from it's lazy flight to take centre stage on a tall Spear Thistle.

THE SECRET TRACK
Page 167
I found this narrow winding track in a lovely quiet corner of Dorset, near Shaftesbury. The high hedge bank created a calm, wind free stretch of pathway. A colourful Tortoiseshell, gliding back and forth, settled down on a patch of thistles. I could see immediately the possibility for a large impressive landscape painting, and began work on it as soon as I got back to my studio.

FLIGHT OVER THE WATER MEADOW, THE ORANGE TIP   Oil   30 × 40 in

Throughout the country, along the edges of fields and footpaths, wildflowers push through a sward of mixed grasses to create a bold display of colour. These lush verges form a natural corridor running between farm fields and linking some key butterfly locations. On some ancient tracks the flowering plants are thick and well established. This is one such track, the Clarendon Way, which runs between Salisbury and Winchester. The reddish purple Knapweed often grows alongside the purple blue Scabious. Both are key nectar sources for passing butterflies, like the Small Tortoiseshell.

THE WILDFLOWER TRACK   Oil   18 × 24 in

BROWN ARGUS AND BEE ORCHIDS

Grazed downland is a flower rich habitat, allowing unusual and specialised plants to flourish. Many orchids fall within this category. I associate them with good butterfly places, as they often share similar habitat requirements. Here I've painted a particular favourite, the Bee Orchid. It is less common than the Spotted, Early Purple and Pyramidal Orchids. In this painting a Brown Argus basks in the June sunshine which warms a dry south facing slope.

Oil   12 × 16 in

A walk along the ridge of the Purbeck Hills is a breathtaking experience with exceptional panoramic views over the Dorset countryside. In this painting, I've tried to capture the feeling of open space, where views from Stonehill Down, stretch out to Brownsea Island and beyond. Here, the rare and localised Adonis Blue is still found in reasonable numbers on a few chalk slopes and sheltered hollows. It flies in August and September on dry sun baked areas, where sheep and rabbit grazing has kept the turf extremely short.

ABOVE POOLE HARBOUR, THE ADONIS BLUE　　　　Oil　　18 × 24 in

TWO TORTOISESHELLS Oil 18 × 26 in

SCABIOUS VISITOR, THE MEADOW BROWN   Oil   16 × 20 in

BLUE BUTTERFLY, MARTIN DOWN

The intense colour of the "Blues" is rarely encountered in nature and makes an almost electric contrast with the landscape. The Common Blue is a softer violet tone than the Adonis, and is painted here on a wonderful stretch of wild grassland, Martin Down. On this National Nature Reserve, managed sheep grazing has created a mosaic of long and short turf areas to encourage an exceptional variety of plant and animal species.

Oil   20 × 30 in

HEATHER VISITOR, THE GATEKEEPER　　　　　Oil　　12 × 16 in

THE OLD TREE STUMP — Oil 18 × 24 in

FLASH OF BRIGHT COLOUR, THE SMALL COPPER

Oil   12 × 24 in

LATE SUMMER BUTTERFLIES — Oil 10 × 12 in

WILD MEADOWS AND GRASSLAND

WAITING FOR THE SUN   Oil   10 × 12 in

Houghton Meadows, Cambridgeshire, is an S.S.S.I. reserve, a remnant of rich wild flower grassland, unploughed since medieval times. I've painted these sheltered fields at their flowering peak, just before the annual cut. Being on damp soil, the plants grow lush and green, with records of over 140 species. Flower-rich meadows were once a common sight and gave butterflies a chance to thrive on farmland. Nowadays, in an age of intensive farming, they are a rare sight.

HOUGHTON MEADOWS NATURE RESERVE   Oil   18 × 24 in

THE HAY MEADOW

Oil   18 × 24 in

Members of the "Vanessids" and the "Whites" are wanderers and will be found in any habitat where attractive nectar sources are available. Hay meadows and set-aside farm fields provide the wildflowers for these passing butterflies.
The peacock is a bold eye-catching species and never fails to gain attention. When resting with wings closed , it will deter predators by flashing it's wings open and shut, revealing it's vivid eye markings.

Oil   24 × 36 in

THISTLE FIELD

    The British countryside would be a very different place without thistles. Butterflies love their purple nectar-rich flowers. They spring up virtually everywhere, in fields, roadsides, hedgerows and along woodland tracks. Different species are well adapted to a wide range of habitats, all producing downy seed heads, which are dispersed by the wind. In "Thistle Field" I was inspired to paint a field corner, thick with Creeping Thistle, always a great favourite with Vanessids.

Acrylic   20 × 16 in

NATURE'S GARDEN, SMALL WHITE AND COMMON BLUE

THISTLE VISITORS

Oil 18 × 24 in

GRASSLAND BUTTERFLIES

Oil 18 × 24 in

By mid June the dusky Meadow Brown begins to appear on every stretch of wild grassland. Here I've painted a typical sheltered hillside where some common butterfly species have found a retreat from the farmland below. A Small tortoiseshell has settled high on a creeping thistle while the browns nectar upon Hawkweed. A Common Blue sits with wings outstretched on it's caterpillars' food plant, Bird's Foot Trefoil.

Oil 12 × 10 in

THE AUTUMN GARDEN

GARDEN VISITORS

Oil   8 × 10 in

MICHAELMAS DAISIES AND RED ADMIRAL

BUDDLEIA LOVERS Oil   10 × 12 in

AGAPANTHUS AND SMALL WHITE Oil 7 × 11 in

MIDSUMMER TORTOISESHELL Oil 10 × 14 in

VANESSIDS ON YELLOW BUDDLEIA Oil 12 × 14 in

Oil 10 × 8 in

RUDBECKIAS AND SMALL WHITE

THE WHITE GARDEN　　　　Oil　　30 × 30 in

VERY RARE MIGRANTS.

Camberwell Beauty

Queen of Spain Fritillary.

Short Tailed Blue.

Monarch.

Long Tailed Blue.

Pale and Berger's Clouded Yellows.

Bath White.

EXTINCT BRITISH SPECIES.

Camberwell Beauty   Large and Beautiful velvet winged migrant from Scandinavia. Most records are from the eastern counties of England and Scotland.

Queen of Spain Fritillary   Intricate, rather angular wings with silver spangled underside, migrates from Southern Europe.

Monarch   Larger than any British resident. Known to travel distances of over 1,500 kilometres. Thought to migrate from America but it is also very popular in butterfly houses and occasional escapes are likely.

Short-tailed Blue   One of the rarest, a wanderer from it's home in central Europe. Only sightings- Channel Islands and the South coast.

Long-tailed Blue   Migrant from the Mediterranean. Very occasional on the South Coast

Pale and Berger's Clouded Yellow   Two almost identical Butterflies, both from southern Europe. Much paler than the Clouded Yellow, the male is a pale lemon yellow, the female is almost white.

Bath White   Has lovely dark blotches, most records are from the south coast in late summer .

Large Tortoiseshell   Extinct around 1980. Lives in N. Africa, Asia and Europe.

Mazarine Blue   Extinct by 1904. Still widespread in Europe

Large Copper   Extinct by 1864. Reintroduction attempts have failed, but further attempts are being planned for the Norfolk Broads.

Black-veined White   Extinct around 1920. Last stronghold was in Kent

Species Index

Adonis Blue	94
Black Hairstreak	78
Brimstone	68
Brown Argus	88
Brown Hairstreak	74
Chalkhill Blue	96
Chequered Skipper	36
Clouded Yellow	66
Comma	120
Common Blue	92
Dark Green Fritillary	128
Dingy Skipper	48
Duke of Burgundy Fritillary	104
Essex Skipper	40
Gatekeeper	146
Glanville Fritillary	138
Grayling	156
Green Hairstreak	80
Green-veined White	58
Grizzled Skipper	50
Heath Fritillary	136
High Brown Fritillary	130
Holly Blue	98
Large Blue (reintroduced)	102
Large Heath	154
Large Skipper	44
Large White	60
Lulworth Skipper	42
Marbled White	148
Marsh Fritillary	134
Meadow Brown	144
Mountain Ringlet	162
Northern Brown Argus	90
Orange Tip	64
Painted Lady	112
Peacock	118
Pearl-Bordered Fritillary	124
Purple Emperor	108
Purple Hairstreak	72
Real's Wood White	62
Red Admiral	114
Ringlet	150
Scotch Argus	160
Silver-spotted Skipper	46
Silver-studded Blue	86
Silver-washed Fritillary	132
Small Blue	100
Small Copper	84
Small Heath	152
Small Pearl-bordered Fritillary	126
Small Skipper	38
Small Tortoiseshell	116
Small White	56
Speckled Wood	142
Swallowtail	54
Wall	158
White Admiral	110
White-letter Hairstreak	76
Wood White	62

Very Rare Migrants

Bath White	204
Berger's Clouded Yellow	204
Camberwell Beauty	204
Long-tailed Blue	204
Monarch	204
Pale Clouded Yellow	204
Queen of Spain Fritillary	204
Short-tailed Blue	204

Extinct British Species

Black-veined White	205
Large Copper	205
Large Tortoiseshell	205
Mazarine Blue	205

OIL SKETCH - GATEKEEPERS

FIELD SKETCH - THE SPECKLED WOOD

FIELD SKETCH - THE SMALL HEATH

FIELD SKETCH - SMALL COPPERS